© Deborah Feingold 2001

Delia Ephron and her dog Daisy

About the Author

DELIA EPHRON is the author of many books of fiction and nonfiction for adults and children, including her most recent novels, *Hanging Up* and *Big City Eyes.* She is also a screenwriter and producer. Her credits include *Sleepless in Seattle, You've Got Mail,* and *Michael.* Her humor, essays, and commentary have appeared in *The New York Times Book Review,* the *New York Times, Vogue,* and *Rosie.* She lives in New York City with her husband, writer Jerome Kass.

How to Eat Like a Child

How to Eat

Like a Child

And Other Lessons in Not Being a Grown-up

DELIA EPHRON

With drawings by Edward Koren

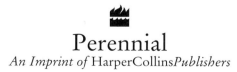

Perennial
An Imprint of HarperCollinsPublishers

First Perennial edition published 2001.

Library of Congress Cataloging-in-Publication Data
Ephron, Delia.
    How to eat like a child: and other lessons in not being
a grown-up / Delia Ephron ; with drawings by Edward Koren.
      p. cm.
Originally published: New York : Viking, 1977.
ISBN 0-06-093675-4
    1. Children—Humor.   I. Koren, Edward.   II. Title.
PN6231.C32 E6 2001
818'.5407—dc21
                                    2001039186

01 02 03 04 05 FG/RRD 10 9 8 7 6 5 4 3 2 1

For Nora, Hallie, Amy, and Lorrie

Acknowledgments

Many thanks and appreciation for their support and editorial advice to Glenn Collins, Ann Banks, Phyllis Henrici, Nancy Evans, Edward Koren, Charles Kaufman, Darrell Brock, and my father, Henry Ephron. My agent, Betty Anne Clarke, was wonderful as ever. I would also like to thank my editor, Rebecca Singleton.

Contents

Introduction   xi

How to Eat Like a Child   1

How to Watch Television   6

How to Watch More Television   7

How to Hang Up the Telephone   9

How to Play   10

How to Laugh Hysterically   16

How to Care for a Pet   19

How to Express an Opinion   23

How to Have a Birthday Party   25

How to Tell a Joke   32

How to Behave at School   33

How to Talk on the Telephone   48

How to Wait   51

How to Act after Being Sent to Your Room   57

How to Celebrate Christmas   59

How to Torture Your Sister   69

How to Ride in a Car   73

How to Sleep   81

How to Say Your Prayers   87

Introduction

One evening I was eating chocolate pudding my way, and I noticed that my friend was eating it her way. It was old-fashioned pudding, the kind you cook. The recipe on the package says, Mix with milk, bring to boil, pour in cups, put in refrigerator. I mention this because only cooked pudding has a skin on top, and, to eat it, I'd made a little hole in the skin and was scooping out the pudding from underneath. My friend had peeled the skin back entirely.

The next day I sat down and wrote something called "How to Eat Like a Child"—a deadpan description of how children eat food. I was a novice. I'd had one article published in a national magazine, *Glamour*, about taking vacations with men. I gave "How to Eat Like a Child" to my friend Edward Koren. He passed it along to *The New York Times Magazine*. He illustrated it, and the magazine ran it on the back page.

On the Sunday it was published and for several weeks after, my phone rang nonstop. Friends called with congratulations and lots of strangers, too, from all over the country. Viking Press asked me to write a book expanding on the article. *New York* magazine offered me a contract as a contributing editor that included—I was really thrilled about this—health insurance. Classes of elementary school children sent me their versions of my article. Letters flooded the *New York Times*, adults writing pages and pages about how they used to eat, and in some cases still eat,

Mallomars* and Oreos.** A *New York Times* editor told me that the only other subject that got this much mail was Israel.

My life as a writer changed. I was launched as a result of five hundred words about children and food. I was so young, so naïve, and so excited that I was idiotic enough to imagine that everything I wrote would strike a universal nerve.

For years I would meet people who would tell me, "Your article is on my refrigerator." What a divine compliment.

When I had decided to become a writer, I calculated that if I used up all my life savings and lived cheaply, I had two years to become self-supporting. It was about three months into the second year when the article "How to Eat Like a Child" came out. Here I was right on track—this much good fortune was too much. As soon as I received my book contract, I blocked. A fellow writer advised me to sit down at the desk every morning for two hours and every afternoon for two hours. If you never get up during that time, he advised, if you never feed your plants or make tea or do anything, you will eventually write. He was right. It's a great cure for writer's block. I highly recommend it.

How to Eat Like a Child and Other Lessons in Not Being a Grown-up was published in the fall of 1978 and became a bestseller. It

---

*Mallomars: Bite off graham cracker bottom, remove chocolate from marshmallow piece by piece, mush marshmallow into ball, pop in mouth.

**Oreos: Split in half/lick off frosting, split in half/lick off frosting, split in half/lick off frosting.

was adapted for television as an NBC special, and became a musical theater revue for children. The revue—book, music, and lyrics—can be obtained through Samuel French Publishing. It has been produced in more than a thousand schools and community theaters in the United States and Canada.

I had thought I would become a journalist, but, when I wrote "How to Eat Like a Child," I found my voice. I learned that I was funny, even wicked. I learned that children and childhood were my emotional landscape. Since then I have written novels, essays, and screenplays on many subjects, but I always feel most at home when I write about kids.

I know I wrote this book, but it feels more like it happened to me. The experience of writing it was so terrifying and having it appreciated was wondrous, even unreal. In the land of publishing, the life span of a book can be very short. To have it republished again more than twenty years later, in its original size and design with Edward Koren's wonderful drawings, is a great joy.

How to Eat Like a Child

How to Eat Like a Child

Peas: Mash and flatten into thin sheet on plate. Press the back of the fork into the peas. Hold fork vertically, prongs up, and lick off peas.

Mashed potatoes: Pat mashed potatoes flat on top. Dig several little depressions. Think of them as ponds or pools. Fill the pools with gravy. With your fork, sculpt rivers between pools and watch the gravy flow between them. Decorate with peas. Do not eat.

Alternative method: Make a large hole in center of mashed potatoes. Pour in ketchup. Stir until potatoes turn pink. Eat as you would peas.

Animal Crackers: Eat each in this order—legs, head, body.

Sandwich: Leave the crusts. If your mother says you have to eat them because that's the best part, stuff the crusts into your pants pocket or between the cushions of the couch.

Spaghetti: Wind too many strands on the fork and make sure at least two strands dangle down. Open your mouth wide and stuff in spaghetti; suck noisily to inhale the dangling strands. Clean plate, ask for seconds, and eat only half. When carrying your plate to the kitchen, hold it tilted so that the remaining spaghetti slides off and onto the floor.

Ice-cream cone: Ask for a double scoop. Knock the top scoop off while walking out the door of the ice-cream parlor. Cry. Lick the remaining scoop slowly so that ice cream melts down the outside of the cone and over your hand. Stop licking when the ice cream is even with the top of the cone. Be sure it is absolutely even. Eat a hole in the bottom of the cone and suck the rest of the ice cream out the bottom. When only the cone remains with ice cream coating the inside, leave on car dashboard.

Ice cream in bowl: Grip spoon upright in fist. Stir ice cream vigorously to make soup. Take a large helping on a spoon, place spoon in mouth, and slowly pull it out, sucking only the top layer of ice cream off. Wave spoon in air. Lick its back. Put in mouth again and suck off some more. Repeat until all ice cream is off spoon and begin again.

Cooked carrots: On way to mouth, drop in lap. Smuggle to garbage in napkin.

Spinach: Divide into little piles. Rearrange into new piles. After five or six maneuvers, sit back and say you are full.

Chocolate-chip cookies: Half-sit, half-lie on the bed, propped up by a pillow. Read a book. Place cookies next to you on the sheet so that crumbs get in the bed. As you eat the cookies, remove each chocolate chip and place it on your stomach. When all the cookies are consumed, eat the chips one by one, allowing two per page.

Milk shake: Bite off one end of the paper covering the straw. Blow through straw to shoot paper across table. Place straw in shake and suck. When the shake just reaches your mouth, place a finger over the top of the straw—the pressure will keep the shake in the straw. Lift straw out of shake, put bottom end in mouth, release finger, and swallow.

Do this until the straw is squished so that you can't suck through it. Ask for another. Open it the same way, but this time shoot the paper at the waitress when she isn't looking. Sip your shake casually—you are just minding your own business—until there is about an inch of shake remaining. Then blow through the straw until bubbles rise to the top of the glass. When your father says he's had just about enough, get a stomachache.

Chewing gum: Remove from mouth and stretch into spaghetti-like strand. Swing like a lasso. Put back in mouth. Pulling out one end and gripping the other end between teeth, have your gum meet your friend's gum and press them together. Think that you have just done something really disgusting.

Baked apple: With your fingers, peel skin off baked apple. Tell your mother you changed your mind, you don't want it. Later, when she is harassed and not paying attention to what she is doing, pick up the naked baked apple and hand it to her.

French fries: Wave one French fry in air for emphasis while you talk. Pretend to conduct orchestra. Then place four fries in your mouth at once and chew. Turn to your sister, open your mouth, and stick out your tongue coated with potatoes. Close mouth and swallow. Smile.

How to Watch Television

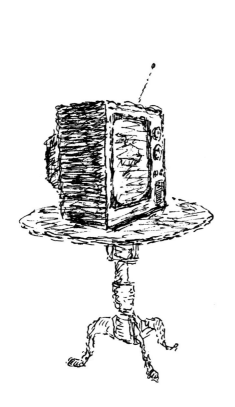

Settle in: Lean back in the chair, throw one leg over the arm and swing it up and down, up and down. As you get really comfortable, exhale out of your mouth several times—directing the air upward to see if you can make the hair of your bangs move in the breeze. Chew on your bottom lip. Stick your finger in your ear and see what you can collect there. Push the dog out of the way with your foot when it blocks your view. Watch the set—eyes wide, mouth slack. Your mother is calling you. Do not hear her, do not hear her, do not hear her.

WHAT TO DO IF THE PROGRAM GETS SCARY

- Go to the bathroom.
- Put a pillow over your head.
- Put fingers in ears, close eyes, and hum.

WHAT TO DO IF THE TELEVISION BREAKS

Turn the channel dial once around fast; turn the set off and on; pull the aerial north, south, east, and west; smash the top of the television with your hand; bang the screen, leaving a greasy handprint across it; say "Fuck," like your dad.

How to Watch More Television

Please, Mom, please. Just this once. I'll only ask once. I promise—if you let me watch this show, I'll go to bed the second it's over. I won't complain. I won't ask for a drink of water. I won't ask for anything. Please. If you let me do this, I'll never ask you for anything again. Never. Please, Mommy, please. You are the nicest mommy. You are the sweetest, nicest mommy. I promise I won't be cranky tomorrow. I promise I'll go to bed tomorrow at nine. Pleasepleaseplease.

Why not! Just give me a reason. I told you I'll be good. I told you I'll go to bed. Don't you believe me? Don't you trust me? Some mom—doesn't even trust her own kid. Look, I'll just close my eyes and listen. I won't even watch it! Oh Mom, why can't I?

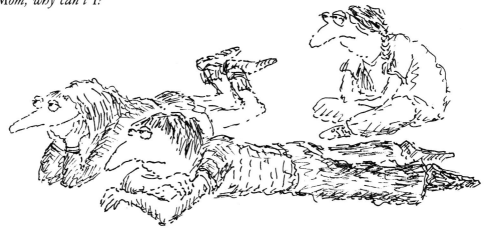

AREA CODE 802

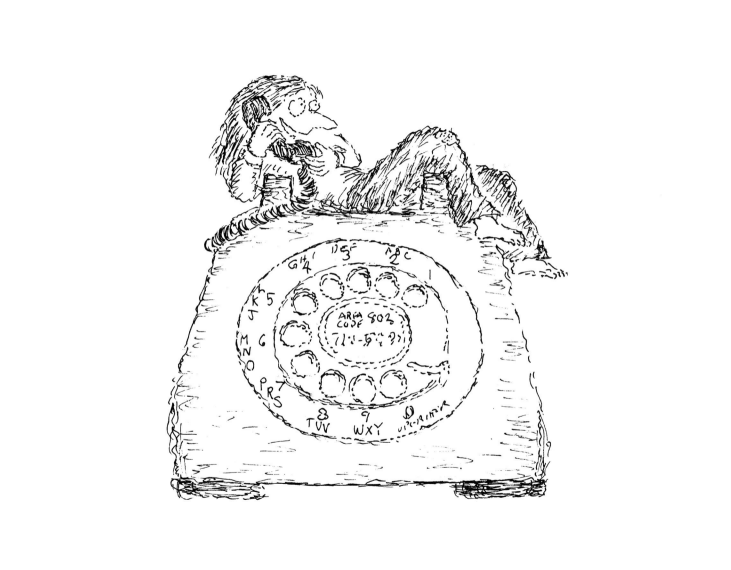

"Good-bye."

" 'Bye."

"Are you still there?"

"Are you?"

"Yeah. Why didn't you hang up?"

"Why didn't you?"

"I was waiting for you."

"I was waiting for *you*. You go first."

"No, you first."

"No, you first."

"No, you first."

"OK, I know. I'll count to three and we'll both hang up at the same time. Ready? One, two, three, 'bye."

" 'Bye." . . .

"Are you still there?"

"Yeah."

"Why didn't you?"

"What do you mean, me?"

"OK, do it again. This time for real. One, two, two and a half, two and three quarters, three. 'Bye."

" 'Bye."

"Hello."

"Hello."

"Are you still there?"

"Yeah."

How to Hang Up the Telephone

How to Play

Wander around the house, trying different seats. Say, "I'm so bored, I've never been so bored, I'm going to die of boredom, there's nothing to do." Your mother will suggest all sorts of games that you might like to play. Say, "Nah, I don't feel like it," to every one. Open the refrigerator, look inside, close the refrigerator.

Ask if you can give the dog a bath.

Ask if you can fly a kite out of your bedroom window.

Ask for the keys to the car—you want to sit in it and listen to the radio.

Decide to make chocolate pudding: Rip open the package so that two-thirds of the pudding mix lands in the saucepan and one-third scatters over the counter. Look at the spilled mix. While you are contemplating what to do about it, use your tongue to check if your tooth is still loose. Then brush the spilled mix off the counter onto the floor and spread it around with your feet so that the mix blends in with the linoleum print. Nobody will notice. Now you are ready to cook: Read the ingredients as

you lick pudding mix off your fingers and, when you come to two cups milk, take a coffee cup out of the cupboard. Use milk for the first cup; to see what happens, substitute water for the second. Turn on the stove and stir. When the pudding doesn't thicken quickly, get bored stirring and conclude that, even though it didn't thicken, it's probably done anyway. Turn off the stove and make imitation braces for your teeth out of aluminum foil. On your way out of the kitchen, run and try to slide on the mix-covered floor.

Turn on the television. Turn it off, but do not go outdoors and play in the fresh air, as instructed.

Pretend to be blind and walk around the house with your eyes closed and an umbrella for a blind-man's stick.

Look in your mother's dresser drawers.

Open the refrigerator, look inside, close the refrigerator. Then, sitting at the breakfast table, make a list of friends in order of favorites. Revise list. Revise again, crumpling up the paper containing the previous lists and flicking it around with your fingers until it falls on the floor. When your mother asks what it is doing there, say that you made a toy for the cat. Decide to form a club.

Call up your best friend and invite him over to discuss details. Agree to make membership cards and have dues. Appoint yourself President. Think about the kids in the class who you do not want to be in the club because they have no personality at all. Call them up on the telephone and announce that you are organizing a club and they can't join.

Stuff a pack of matches in your pocket.

Yell, "Hey Mom, watch me, watch me, watch me, watch me," and stand on your head. Repeat six times.

As soon as your friend arrives to play with you, dig to China. If your friend says that he dug past clay, insist that you heard voices.

Play Clue. Argue over who gets to be Colonel Mustard.

Play Monopoly. Argue over who just went. Argue over whether it was a five or a seven. Argue over whether it counts if the dice are off the board. Argue over whether you counted from Marvin Gardens or Water Works. Argue over whether you skipped one when you counted. Argue over whether you counted right when you landed on Free Parking. Argue over whether you already collected two hundred dollars when you passed Go. Do not finish the game.

Play fish. Check out your friend's hand when he carelessly holds his cards down.

Play war. Sit on an ace.

Ask if your friend wants to play fifty-two pickup and throw the cards in the air.

Play checkers. Say, "I want the reds." Say, "I want to take it over." Say, "Just a second—don't do anything till I get back," run into the kitchen, open the refrigerator, take a swallow of milk directly from the carton, and return. Say, "The move doesn't count till I take my hand off the checker."

Tap a tune on the table and see if your friend can name it. Ask him, "Wanna stay for dinner—we're having spaghetti."

Throw raisins into the air and try to catch them in your mouth. Miss. Throw raisins into your friend's mouth while he throws them into yours. Miss. Feed the dog raisins. Lie on the ground and put raisins on your face and let the dog lick them off until your mother comes out and says, "Don't let the dog lick your face, he has germs." Walk over to your friend and casually pat him on the back. Say, "So how're you doing?" while you sneak a few raisins under his collar and down his shirt.

Play school. Take turns being the teacher and sending each other to stand in the corner.

After using your bed as a trampoline, transform your room into a giant spiderweb, running lines and lines of kite string from bedposts to lamps to table and desk legs to bureau handles and doorknobs. Think that the web looks really neat, as you crawl out of the room underneath it, but do not show it to your mother.

Announce that you are double-jointed. Prove it by pulling your thumb back until it touches your wrist. When your friend says, "Me too," and presses down the fingers of one hand backward, show off your legs: Sit on the floor with them bent at the knee to a hairpin angle. Let your friend admire you in this position. Then take turns trying to smile with one side of your mouth and frown with the other. Lie on the floor and balance a penny on your nose. See who can hold his breath the longest. In order to win, cheat: Keep your cheeks puffed out as evidence that you are holding your breath, but actually breathe through your nose. If your friend accuses you of faking, deny it. Then say, "Look what I can do," and demonstrate: Crouch down like a catcher, put your head between your knees, breathe in and out quickly sixty times, stand up, and pass out.

There is the doorbell—your friend's father has arrived to pick him up. Overhear his dad ask your mom, "Did he behave himself?"

Sit at the table drinking juice. Hold glass in front of mouth like a microphone and deliver a broadcast: "Well, folks, it's Saturday at the house and there's not much happening. I'm sitting at the table drinking juice. On the news front, it looks like my dad may go to Columbus tomorrow if the weather holds, and we'll have more on this later. The home team, fighting from behind, won at Clue, war, and fish. The results are not in yet on Monopoly. We'll keep you posted on this. And now a word from a guest in our studio audience."

Put the microphone in front of the dog's mouth as your mother walks in. She will ask if you have been looking in her dresser drawers. Say, "No." If she says, "I don't mind if you did it, I only mind if you lie to me," say, "No, Mom, honest." Agree to dismantle the spiderweb instantly, and when she informs you that the chocolate pudding didn't jell, tell her that's the way you wanted it. Then, to the glass, say, "Roger. Over and out." Slide down, leaving your chair like a snake, and exit from under the table.

How to Laugh Hysterically

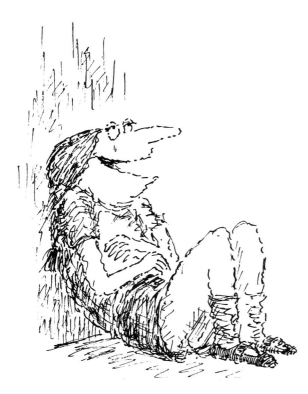

Call up a Chinese restaurant and have a Chinese dinner sent to your teacher's house.

Call up a pizza parlor and send your teacher seven pizzas.

Call up a diaper service and sign up the lady next door for delivery.

Call up Kentucky Fried Chicken. Say, "How large are your breasts?"

Call up a stranger. Tell him that you are the telephone company repair person. You are working on the line. If he answers the telephone within the next ten minutes, he will electrocute you. Call back and let the phone ring and ring until he finally answers it. Scream.

Call up a boy in the class and don't say anything.

How to Care for a Pet

Turtles: Give them two names that go together. Suggestions: Wienie and Bun, Sunshine and Moonglow, Bacon and Egg. Lie on the floor and let the turtles walk on your stomach. Introduce them to each other. Examine the bottom of a turtle to see how it pees. Put it down on its back to see if it can turn itself over. Do not notice when a turtle disappears under the bed.

You now have one turtle living in the plastic turtle container with a spiral ramp for it to exercise on. After painting its shell with nail polish, forget to feed it.

Wounded bird: Take home, put in shoebox lined with shredded newspapers and cotton, and feed with an eyedropper. Call up the SPCA. Worry about whether the other birds will accept it when it returns to the wild.

Dead bird: Bury in shoebox. Say prayer. Mark grave with rock.
Alternative method: Put in desk drawer and forget it until it smells.

Goldfish: Overfeed. Flush down toilet.

Daddy Longlegs: Pull off the legs and watch them move around by themselves.

Snake: Know that it is somewhere in the house, but have no idea where.

Alligator: First day: Bring it home and let it play in the bathtub. Show it to your mother. Put in jar for night.

  Second day: Return home from school and find the jar empty. Tell your mother that the alligator has disappeared. She will say that she can't imagine what happened to it.

Frog: Cage in empty mayonnaise jar with holes punched in top for air. Stick in a few leaves so that the frog has a natural environment. Carry around jar all day. Place it next to you on the table while you eat lunch. Watch the frog jump and bang its head on the glass. Drop in a piece of hot dog. Ask Mom how soon Dad will be home. Turn on the TV and at each commercial, yell, "Hey Mom, how much longer?" As soon as you hear Dad at the front door, yell, "Hey Mom, Dad's home." Show him the frog, let him admire it, and listen to him say the same thing that he said about the ladybug and the caterpillar: "If you were a frog, would you like to live in a jar?"

Gerbil: Ask the pet shop for two males. Wake up one morning and find six gerbils. Come home after school and find one gerbil. Scream.

Rabbit: Pick it up. Put it down. Pick it up. Put it down. Pick it up. Put it down. Pick it up. Put it down. Pick it up. "Would you please leave that poor rabbit in peace?" Put it down.

Dog: Tell your parents that you want a dog more than anything in this world. Promise that you'll take care of it. Cross your heart and hope to die. They won't have to do a thing. You'll walk it and feed it. Please. Please. Pretty-please. Pretty-please with sugar on top. Pretty-please with whipped cream and a cherry. Please, Mom, please. You are, too, old enough. When they say that you have to wait one more year, stamp your foot. Scream, "You never trust me; you never believe me. Why don't you trust me?"

Run to your room, slam the door, open the door, and yell, "It's not fair." Slam the door again. When your mother comes to your room, have the following conversation:

"That's enough," says she.

Say, "All right," as though it isn't.

"I said, that's enough."

"All I said was 'All right.'"

"It's not what you said, it's how you said it."

"OK, Mom, but . . ."

and repeat entire scene from "I want a dog more than anything in the world" to "Why don't you trust me?" Convince her.

When your mother says that it's time to feed the dog, say, "Just a second." When she says it's time to walk the dog, say, "In a minute." After she reminds you again, tell her that you just want to watch the end of the TV show. Then take out the dog, complaining that it's so boring to walk it and besides you can't find the leash. Use your belt instead.

Each day, procrastinate and complain until your mother finds it easier to feed it and walk it herself.

Kitten: Pick up by tail. Pick up by scruff of neck. When your dad says, "Don't you think we ought to leave the kitten alone for a while?" put it down. When your younger brother comes in and picks it up by the rear end, grab it by the front end and try to get it away from him. Scream, "You're supposed to leave the cat alone." Keep pulling until your dad returns and says, "Didn't you hear? Leave the cat alone." Glare at your brother. Hiss, "Jerk. Now look what you've done." Hit your brother and explain that you didn't mean to, your hand slipped. As soon as your brother and dad leave the room, say, "Here kitty, kitty, kitty." Pet and hug it. Then, after checking if the coast is clear, drop the kitten over the banister down to the first floor to see if, when cats fall, they always land on their feet.

How to Express an Opinion

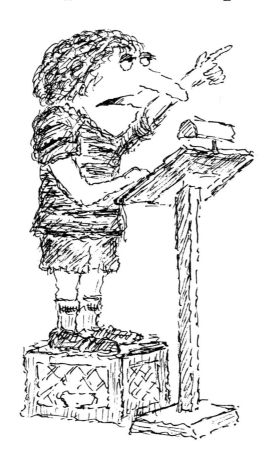

Yucky
Gross
Dis–gusting
Ugh
Sick
Sickening
Scuzzy
Smell-y
Oh, barf
Creepy
Icky
Obnoxious
Boy, is this dumb
Creeps
Crummy
Vomitrocious

23

How to Have a Birthday Party

INSTRUCTIONS FOR THE BIRTHDAY GIRL

Rearrange the place cards.

After you are dressed and ready two hours before the party begins, check them again to reassure yourself—you are seated at the head of the table, your absolute best friend is next to you, and your little brother is as far away as possible. Wish that he didn't have to be there, period. Look at the clock.

Go to the window. See if anyone's arriving. Go to the kitchen. Look at the bakery box on top of the refrigerator and wonder what the birthday cake looks like. Open the refrigerator. Close it. Run to the bedroom. Run to the living room. Ask how much longer. Tell your mom how excited you are. Twirl. Ask if you can play hot potato first, musical chairs second, and pin-the-tail-on-the-donkey third. Look at the prizes. Ask if you are allowed to win a prize even though it's your party. Ask how much longer. Run to the window and see if anyone's arriving. Turn on the TV. Turn it off. Return to the party table. Trade your red favors for turquoise. Trade back. Run to the window. Yell, "Someone's here, someone's here, someone's here. Mom, Dad, someone's here." Run to the door and open it before they get out of the car.

When guests arrive, greet them with, "What did you bring me?" and grab the present out of their hands.

INSTRUCTIONS FOR THE GUEST

Say, "Happy birthday." Surrender present.

Stand on a chair and yell, "Guess who's sitting next to me?" Call out the name of your best friend. Announce that the cake will be chocolate. Whisper in the corner with your best friend and shoo away guests who try to hear. Tell her what the prizes are. Discuss whether one of the other guests is obnoxious. Walk around the room with your arms around each other.

When two guests prefer to talk to each other rather than watch you open presents, ask them to stop. Read gift cards loudly to attract them. Pass the cards around. Notice that each time you unwrap a present, those two kids are not paying attention to you. Finally, when one of the other guests wanders out of the room, ask to speak to your mother—privately. Tell her that the guests are not acting the way they're supposed to. Tell her that they're ruining the party. Tell her they won't be quiet and watch. Ask her to make them.

Want to die when the birthday girl reads your card aloud, and look around to see if people are listening. Tell a few guests that they're not supposed to talk while the presents are being unwrapped. As she pulls off the ribbon, say, "You're going to hate my present. I know it, you're going to hate it." Do not take your eyes off the present. Beam. The minute it comes into view, yell what it is. Tell her that it's real expensive—ten dollars. Say that there was a cheaper kind, but your mom said it wouldn't last. Tell her that it came in green, but you thought red was better. Tell her the name of the store where you bought it and say, "Wanna see how it works? Here's how it works." Grab it and demonstrate. As soon as the birthday girl begins on the next present, lose interest. Wander out of the room to check out the party favors.

When a present turns out to be something that you already own, say so.

When a present turns out to be identical to another, try to be tactful: Say, "Oh, no!" and smash the palm of your hand into your forehead.

When a present is revealed to be a book, think, "Yuck."

Move your place card. Switch your party favors. Sit down at the table and raise your hand as a person who prefers milk to Hawaiian Punch.

Blow out the candles on your cake with the carousel theme while your parents assist by blowing over your shoulder. Wish for a horse. Forget that if you talk before cutting a slice your wish won't come true and yell, "Who wants a rose on their slice? Raise your hand." If someone reminds you about not talking and the wish, think that the talking doesn't count because it was only one sentence. Lick the frosting off the candles after picking them off the cake.

After singing a verse of "Happy Birthday to You," sing, "You belong in the zoo, you look like a monkey, and you act like one, too." Yell, "I want a rose," over and over and wave hand in air. When milk arrives, request juice instead.

Empty your bag of party candy on the table. Trade a purple lollipop for a red one, four Tootsie Rolls for a miniature Butterfingers bar, and a plastic ring with a green stone for one with a gold stone. Be unable to pawn off a green lollipop. Do not eat black Necco wafers.

Ignore what your little brother is shouting to you from the other end of the table. Wish that he would disappear.

Stand up. Blow on your party horn. Reach down three seats to grab an extra hat; fight over it; knock over a paper cup filled with Hawaiian Punch. Sit down. Blow on the party favor that uncurls and stretches out so that it jams into your neighbor's ear. Do it again. Shriek. Stand up. Blow into the bag that once contained candy and, holding the opening closed with one hand, smash the bag with the other hand to pop it. Look around quickly so that other guests know it was you; sit down quickly so that adults don't see that it was you.

Stand up and blow on your party horn. Sit down. Blow on the favor that uncurls and stretches out so that it jams into your neighbor's ear. Shriek. Blow on the favor again. Stand up. Blow on the horn. Sit down. Fail to get the attention of your guests by banging your spoon on the table and screaming, "Order in the court. Order in the court." Eat two bites of cake and leave the rest.

Mush your ice cream while you announce that there were better favors at the last party. When an adult walks by, ask for another bag for your candy: Yours got ripped. Eat two bites of cake and leave the rest.

Do not want to play party games. Want to run them. Shout that everyone should get in line. Tell them to be quiet, and if they talk anyway, say that there are good prizes and that they had better shut up if they want some. Find a reason to make all the guests raise their hands. Tell a girl who refuses to play pin-the-tail-on-the-donkey that she has to. Reason with her, explaining, "Everyone has to play." Then, when she still refuses, do not know what to do. Show her to your mother.

Your mother will suggest to her, "Wouldn't you just like to try the game?" and when she says no, your mother will explain to you that since your guest really doesn't want to play, she shouldn't have to. You do not know what to do. She's ruining your party!

Everyone's ruining it!

Burst into tears.

Run to your room.

As soon as your mother comes after you, tell her everything while crying and hiccuping. The party never goes right, never! It's never the way you want it! Why won't the guests behave? Why? Don't they understand that the party won't work unless they do what *you* tell them? "I'm never inviting that girl to another one of my birthdays as long as I live!" Allow your mother to convince you to return.

After losing musical chairs and hot potato, refuse to play pin-the-tail-on-the-donkey.

When the birthday girl tells you to play, say that you don't want to. When she pulls on your arm, jerk it away. Say that nobody can make you play if you don't want to. Give in only after the birthday girl cries and her mother begs. Then, when a grown-up blindfolds you for pin-the-tail and asks if you can see, lie. With head tilted back and peeking beneath scarf, walk very crookedly toward donkey, bumping into people on sidelines. Start to pin the tail on the head; then, legs; then, at the last second, pin it on the rear end. Pull off scarf and look surprised.

Stand on a chair to announce the game winners, award prizes, and make a big production out of it.

When you get the prize, think that the second prize for hot potato was better than first prize for pin-the-tail.

Be too busy playing chase to say good-bye to guests.

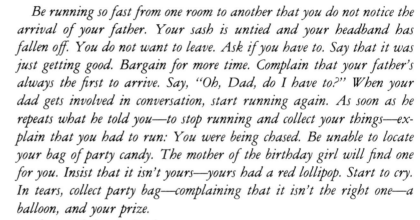

Be running so fast from one room to another that you do not notice the arrival of your father. Your sash is untied and your headband has fallen off. You do not want to leave. Ask if you have to. Say that it was just getting good. Bargain for more time. Complain that your father's always the first to arrive. Say, "Oh, Dad, do I have to?" When your dad gets involved in conversation, start running again. As soon as he repeats what he told you—to stop running and collect your things—explain that you had to run: You were being chased. Be unable to locate your bag of party candy. The mother of the birthday girl will find one for you. Insist that it isn't yours—yours had a red lollipop. Start to cry. In tears, collect party bag—complaining that it isn't the right one—a balloon, and your prize.

If reminded, say thank you.

Go home.

Throw up.

How to Tell a Joke

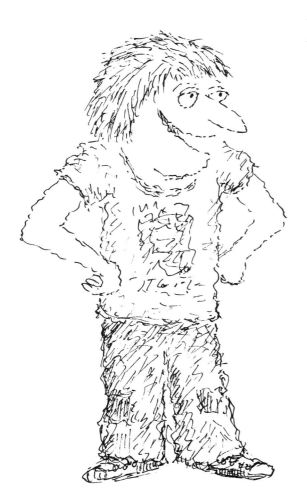

"Wanna hear a dirty joke? A pig fell in the mud. Get it?"
Immediately repeat ten times.

How to Behave at School

Ma, I don't feel good. Maybe I shouldn't get up today. I feel sorta blah. I don't know—I just feel yucky all over. Ma? Ma, would you feel my forehead? I don't? Are you sure? Are you positive? OK, I'll get up. I'll get up, but you'll see—I'll probably just get to school and have to turn around and come home again.

Arrive at school late. Explain that you are tardy because you couldn't find your shoe.

As soon as the teacher turns to write on the blackboard, open your desk, pull out *Mad* magazine, and put it inside the language arts workbook. Read *Mad* magazine while it looks as if you are reading language arts.

Chew a pencil, tear off the corner of a piece of paper, and write: "You are a dodo. Pass it on." Gripping the pencil between your teeth like a pirate with a knife in his mouth, fold note in half, quarters, eighths. Use your ear as a pencil holder as you drop the note on the floor and pass it with your foot to your friend across the aisle. Then, pretending to play the drums, tap your desk with a pencil, and when it's time to hit the cymbals, tap the head of the kid in front. If he turns around and says, "Cut it out," say, "Cut

what out?" As soon as he faces front again, kick him and say, "Sorry, I didn't mean to." Click your pen.

Deny that you are chewing gum and stick it on the roof of your mouth.

Whisper. Stop when the teacher asks if you'd prefer to spend class in the hall. Ask to change your seat.

Pretend that your pencils are ships; steer them around your desk and make them collide. Look at the clock.

When the teacher asks for a volunteer to take names while she is out of the classroom, raise your hand, shake it frantically, stretch so that your body is nearly a horizontal line between your desk and the teacher, and call out, "Me, me, me, me, me, me, me." You do not get chosen.

What to Do While the Teacher Is Out of the Classroom: Hold your nose and say in a high-pitched voice, "Now class, behave." Run to the front of the room and draw your fingernails down the blackboard. Return to your seat like Groucho Marx, hunched over, looking both ways, wiggling eyebrows, and chomping on a pencil as if it were a cigar. Get your name taken.

Get it erased by threatening to get the name-taker at recess.

Sail a paper airplane and when it lands, raise your hands, clasp them above one shoulder, then the other: You are the champ. Get your name taken.

Throw an eraser.

When a kid shoots a spitball and doesn't get his name taken, say, "How come you took my name and not his?" Get his name taken while he tells you to shut up, calls you by your last name only, and says, "Mind your own beeswax." Burp.

Smell something funny. Shriek, "Someone laid one, someone laid one, silent-but-deadly, smell-y, whew, P.U., major fart alert, major fart alert, major fart alert." Wave hand in front of face. Crack up. Pound desk. Hold nose while each kid in the class also holds his nose and insists that it was another kid.

Yell, "She's coming," and fall out of your chair just as the teacher returns.

Ask to get something from your coat in the cloakroom.

Ask to sharpen your pencil.

Tell the teacher, for the second time this week, that you do not have your homework because the dog ate it. She will say that if this kind of behavior continues, she will have to note it on your permanent record card.

Look at the clock.

Ask to get a drink of water.

IN THE HALL

Look in all the classrooms you pass, stopping at one or two long enough to attract attention and distract the students—stick thumbs in ears and wave fingers, or scratch armpits like a monkey and heave up and down. If a class has its door closed, jump up to see through the window on the door. Play hopscotch, using floor tiles as squares. Stand against the wall and inch your way down

HOW TO ACT IF YOU DO NOT WANT
TO BE CALLED ON

• Make yourself invisible. Align head and shoulders with those of a student directly between the teacher and you. If the teacher moves, adjust alignment.

• Make yourself inconspicuous. To accomplish this, assume a casual pose. Concentrate on fitting the top of the pen into the bottom; perhaps even hum to yourself. Or engage in nonchalant play with a pencil: Hold it upright, point against paper, and slide fingers from eraser to tip. Turn pencil over; slide from tip to eraser. Turn and slide. Turn and slide.

If the teacher calls on you anyway, do not respond immediately in the hope that a kid with the answer will just yell it out. If no one rescues you and the question calls for a yes or no response, pick one. Otherwise, give a joke answer. The class will laugh. The teacher will say that it won't be so funny when you get your report card.

the hall—you are in a spy movie. When you reach a corner, peek around it. After turning up the water at the fountain to see how high it will go, fill up your squirt gun. Walk back to class with the point of the gun in your mouth; keep pulling the trigger.

As soon as you return, check the clock to see how much time you killed.

AT RECESS

With tongue, remove gum from roof of mouth and continue chewing.

Stand around. Discuss bedtimes. Say that you stayed up until midnight to finish your homework. Your friend will answer, that's nothing, he stayed up until two o'clock, and you can respond, "Big deal, last week I was up until four." Add that on weekends you can stay up as late as you want, once you were up all night, and has he ever been to a drive-in movie? Your dad took you. Show each other your fillings. Announce whether you are a Democrat or a Republican and take a position on the coming election based on your parents' conversation the night before. Discuss Hydrox versus Oreo. Compare chunky peanut butter and creamy. Argue about the most sickening vegetable. Compare number of rooms in your house, counting closets. Compare allowances. Choose whether you would rather freeze to death or be burned alive. Feel each other's muscles. Say that you got a break on your new car 'cause your dad knows the dealer. Say that your dad thinks foreign cars are better. Say that he likes "four on the floor." Discuss brands of sneakers. Try to step on each other's shoes. Demonstrate how to make a fart noise by putting your hand in your armpit and squeezing. Say that you know the longest word in the English language—antidisestablishmentarianism. Bet that no one can spell it.

Walk around wearing wax mustaches and red lips.

See if M&Ms will melt in your hand.

Eat red-hots and show off your red tongue.

Promise that you'll be a kid's best friend if he gives you a Tootsie Pop. Wish that you would hurry up and get to the center of it while you listen to a riddle: Why did the little moron take a ladder to the party? Because he heard that drinks were on the house.

Say, "Guess what?" When a kid says, "What?" say, "That's what." Do it again and again until you come to a kid who, when you say, "Guess what?" says, "That's what."

There is a student whose head is kind of flat on top. Skip around the playground chanting, "Flathead, flathead, flathead."

To be immune from a kid with cooties, give yourself a cootie shot.

Put your finger over the spout of the drinking fountain to direct the spray so it hits everyone. When the water soaks a kid's pants, shriek that everyone will think he wet them. Hold the faucet handle for a friend to take a drink and, when he leans over, let the water die down. Say, "OK, OK, I'll do it right. Honest. Word of honor. Trust me. Really, don't worry," and when he leans down again, turn it up full force. Take a drink, hold water in mouth, chase a kid and spray him. Do it again, but this time burst out laughing before you get a chance at a good shot. While screaming, repeat all water activities until the playground supervisor threatens to bench you.

Stand under the jungle gym and look under girls' skirts. Tell a girl that you'll give her a nickel if she'll climb up.

Lie on the ground just in case a girl walks by.

Ask a girl in a dress to stand on her head.

Listen to a kid tell a dirty joke. Laugh hysterically. Do not have any idea what you are laughing at.

Envy the kid with a broken leg. Beg for a turn on his crutches.

Say "Fuck," and see what everyone does.

Pretend that you do not know your little sister.

Be chased by girls.

When a boy falls off the jungle gym and gets his front teeth knocked out, yell, "I'll do it, I'll do it, I'll do it," collect the pieces, and turn them into the nurse's office. Even though the nurse's office smells funny, hang around to try to get a look at the boy.

Play tag. Argue about whether you were touched. Play war. Argue about whether you were killed. Play kickball. Argue about whether you were out. Play dodge ball. Argue about whether you were hit. Play statues. Argue about whether you moved. Take sides when everyone else argues. If necessary, scream about fairness and call the other side names, such as jerks, creeps, crumbs, liars, and dirty cheaters.

Get benched. On your way there, think that it's not fair. Think that the kid you hit, who hit you back and you hit again, will be sorry. Think that you'll get him, that you'll never speak to him again, never walk to school with him again, invite him to your house or birthday party. Think, "Just let him come within ten feet of me."

After sitting on the bench for a minute, see how far you can stand from the bench and still be considered benched. Stick your gum under the bench.

IN THE BOYS' LAVATORY

Aim at the side of the toilet. Do not flush.

HOW TO EAT A CAFETERIA LUNCH

Plan to flip butter off fork to ceiling, where it will stick until the next group comes in to eat; then it will fall on someone's head. Be unable to put this plan into effect.

Shoot off paper, aiming at the cafeteria monitor

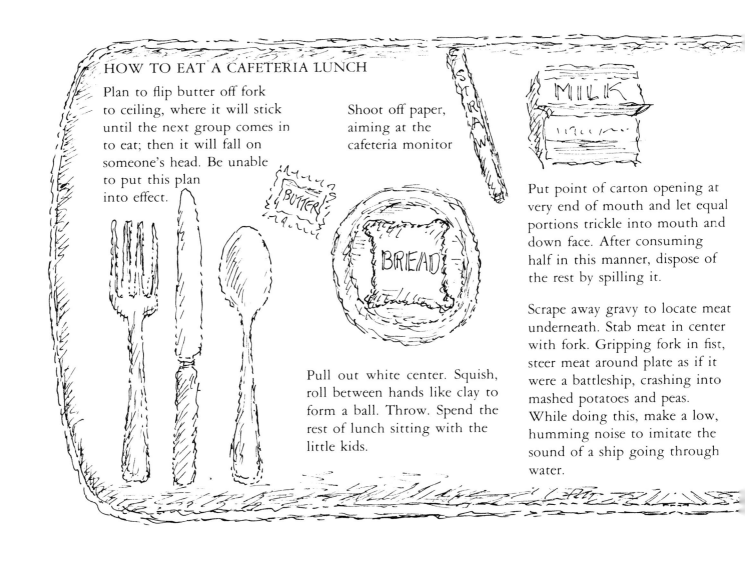

Pull out white center. Squish, roll between hands like clay to form a ball. Throw. Spend the rest of lunch sitting with the little kids.

Put point of carton opening at very end of mouth and let equal portions trickle into mouth and down face. After consuming half in this manner, dispose of the rest by spilling it.

Scrape away gravy to locate meat underneath. Stab meat in center with fork. Gripping fork in fist, steer meat around plate as if it were a battleship, crashing into mashed potatoes and peas. While doing this, make a low, humming noise to imitate the sound of a ship going through water.

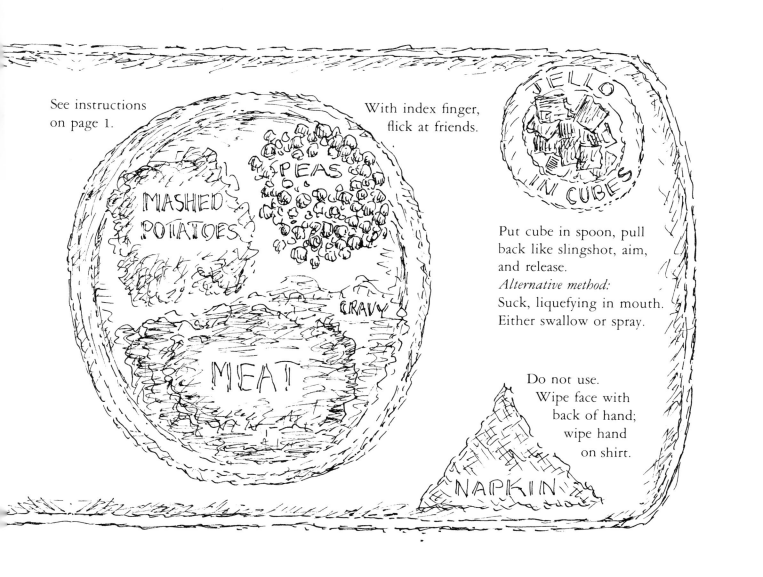

See instructions on page 1.

With index finger, flick at friends.

MASHED POTATOES

PEAS

GRAVY

MEAT

JELLO IN CUBES

Put cube in spoon, pull back like slingshot, aim, and release.
Alternative method:
Suck, liquefying in mouth. Either swallow or spray.

Do not use. Wipe face with back of hand; wipe hand on shirt.

NAPKIN

HOW TO EAT A BAG LUNCH

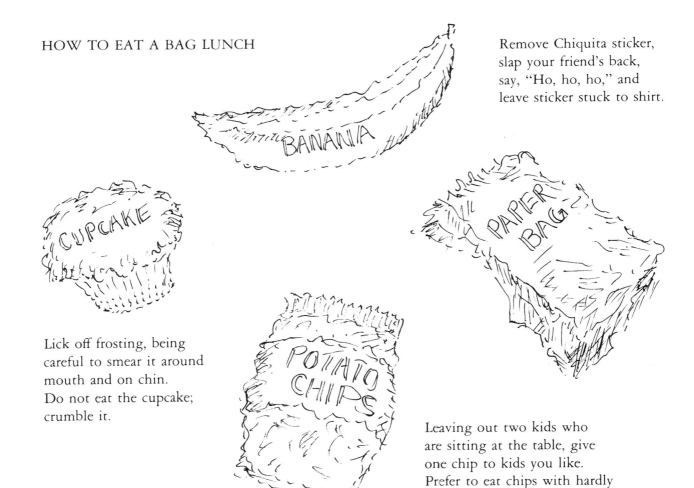

Remove Chiquita sticker,
slap your friend's back,
say, "Ho, ho, ho," and
leave sticker stuck to shirt.

BANANA

CUPCAKE

PAPER BAG

Lick off frosting, being
careful to smear it around
mouth and on chin.
Do not eat the cupcake;
crumble it.

POTATO CHIPS

Leaving out two kids who
are sitting at the table, give
one chip to kids you like.
Prefer to eat chips with hardly
any brown in them.

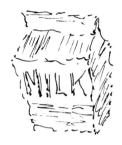

After spilling, throw carton in garbage can as if it were a basketball. Then move sideways on the bench, shoving the person next to you until the kid at the other end falls off.

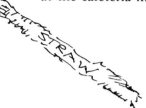

Shoot off paper, aiming at the cafeteria monitor.

Brag about having fried chicken. Say that you had it last night for dinner. Say that you have fried chicken at your house three times a week, easy. Announce your favorite part. Wave the drumstick while chanting, "Roast chicken, boo; fried chicken, yay." Pick off skin with fingers; then consume.

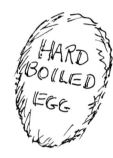

Do not eat, but leave the peeled-off shell and bare egg covered with dirty fingerprints on the cafeteria table.

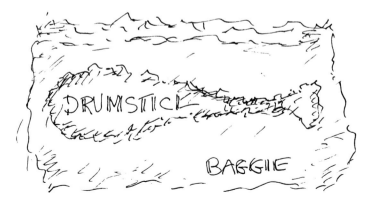

Blow in air, hold closed, and smash. Spend the rest of lunch sitting with the little kids.

AT ASSEMBLY

"I pledge allegiance to the flag of the United States of America. And to the Republic for Richard stands, one nation, under God, indivisible, with liberty and justice for all." Sit down.

Crane neck around to see where teacher is sitting.

As the film on fire prevention—starring Otto, the talking car—begins, take a marble out of your pocket and roll it between your hands. Putting hand up to mouth as if to stifle a yawn, pop the marble in. With your tongue, shove it over to your cheek so that it sticks out and makes a lump. Push it with your tongue over to your other cheek. See how fast you can move it from one cheek to another. Take it out. Clench it in your eye like a monocle. Stuff it in your nose. Leave it there—half in, half out—then poke your best friend. He shouldn't miss it.

Listen to the kid at the end of the row crack his knuckles. Listen to the kid next to him crack his knuckles. Listen to the third kid crack his knuckles. Listen to the fourth kid. Crack your knuckles. Listen to the kid after you.

Fight over the armrest.

Look at the projector to see how much film is left.

Drop your marble.

As soon as the projector breaks, start whispering. Stop when the teacher snaps her fingers, but do not look at her. She might point at you and then stick her thumb over her shoulder: Out! Try to get the attention of your girl friend by enlarging yourself somewhat: Clasp hands behind head and rise up, stretching.

Watch the kid fix the projector. Wish that you were a member of the AV squad.

As soon as the film resumes, fall asleep.

"The assembly is now adjourned. Will the classes please leave quietly, passing out of the auditorium front rows to back."

WALKING HOME

Step on a crack, break your mother's back. Take it one step per pavement square. *Step on a line, break your mother's spine.* Two steps per pavement square. Skip every other square. Pull leaves off hedges. Kick a rock as you walk. Avoid quicksand. Be your own horse: Say "Giddyup," and slap your thigh. Jump up; grab a branch of a tree. Run—you are being chased by red ants. Ring the bell, hit the knocker, slam the door, shout, "Hey Mom, it's me," and race to the bathroom.

"How was school?"

"Fine."

"What's new?"

"Nothing."

"What did you do today?"

"I don't know—the usual."

Then add, "I'm starved," and on your way to the kitchen, tell your mom the riddle: Why did the little moron take a ladder to the party? Because he heard that drinks were on the roof.

How to Talk on the Telephone

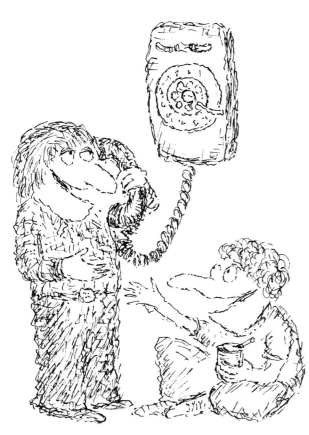

Hello. Is your refrigerator running? Then you'd better go and catch it.

Hello. Is this the cigar store? Do you have Muriel in a box? Well, you'd better let her out before she suffocates.

Hello. Is this the drugstore? Do you have Bayer Aspirin? Cover it up; it's going to catch cold.

Hello. Is this Mr. Goose? Is your mother there?

Hello. I'm taking a survey. Have you ever eaten peanut butter and amatta. What's amatta? I don't know. What's amatta with you?

Hello. Are you English? Are you Swedish? Are you Italian? Are you Finnish? Well, I am. Good-bye.

Hello. I'm taking a survey. Do you have a television? How

many kids do you have? Do you eat spinach? Thank you. Good-bye.

Hello. We have a call for you from Washington. Please stand by; we'll call you back.

Hello. Are you expecting a call from Washington? Well, then you must be crazy because he's been dead for a thousand years.

Hello. Is Ralph Peterson there?
Hello. Is Ralph Peterson there?
Hello. Is Ralph Peterson there?
Hello. Is Ralph Peterson there?
Hello. Is Ralph Peterson there?
Hello. This is Ralph Peterson. Are there any messages for me?

Hello. Is this the grocery store? Do you have pig's feet? Then how do you get your shoes on in the morning?

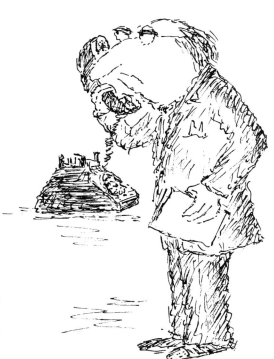

How to Wait

FOR A HAMBURGER

Grip lunch counter and swing from side to side on stool; let go and make one complete spin around. Repeat until your dad makes you stop. Rest elbows on counter, chin in hands, and stare at the milk-shake machine. Wish that you had one at home. When each hamburger appears above the grill, wonder if it is yours. Flip the metal flap covering the spout of the sugar dispenser. Feel under the counter for gum.

IN THE SUPERMARKET PARKING LOT

You are sitting alone in the car. Slide over behind the wheel. Check out the equipment: Turn on the wipers, activate the squirter, turn the wipers off; test the turn signals. Drive: Put both hands on the wheel and turn it back and forth, back and forth, back and forth. Flip on the radio. With your elbow on the window ledge and one hand on the wheel, relax and look around.

TO BE PICKED UP AFTER SCHOOL

"I'll wait one minute and then she'll be here." Say this to your-self and count off seconds—one hippopotamus, two hippopota-mus, three hippopotamus, four hippopotamus. After reaching sixty, say to yourself that she'll be here after the fifth car goes by. Count five. Say with conviction that she'll be here after the light changes twice. Watch it change. Watch kids ride by on bikes. Watch kids get picked up. Is that car yours? Run to the curb. It isn't. Feel embarrassed. Feel conspicuous. Run back to the fence so that anyone watching will think that you were practicing run-ning. Whistle and lean against the fence. Decide that you are thirsty. Decide that you are dying of thirst. Imagine that your mother doesn't arrive for seven days and that you just make it to the car, crawling. Worry that your mother might never come at all.

Put your notebook down on the grass and watch the kids in the playground. Pressing your face against the wire mesh, imagine that you are in prison looking at free people play in the sun.

You have to go to the bathroom. Stand with your legs crossed.

Notice that your teacher is leaving school. Hope that she sees you; hope that she doesn't see you.

Hear a honk, run to the curb, get into the car, and say, "It's about time."

FOR A CHECKUP

After looking at the aquarium, lie on the floor and, with Magic Markers, color the paper that your mother brought along to keep you amused. While you are in the middle of drawing a rising sun, let the pen slip over onto the rug and press the point down into the shag to see if it absorbs the ink. When the carpet turns yellow, move the paper over to hide the spot and continue drawing.

TO LEAVE THE RESTAURANT

Rub your back against the booth to scratch an itch while listening to your parents talk. Think how incredibly boring their conversation is. Covering your face with your hands, pull all your features downward so that you resemble a basset hound, and look around. Wish that the baby crying at the next table would shut up. Think that if the baby were yours, you would stuff a napkin in its mouth. Ask to go to the bathroom.

Take the long way, cruising past the jukebox and checking out the other diners. As soon as you arrive in the bathroom, collect fifteen soap packages from the soap dispenser and cram them in your pockets. Turn on all the faucets. Turn them off. Look in the mirror and smile at yourself. The bathroom smells funny: To try to avoid the odor, breathe through your mouth and not your nose. Observe the Kotex machine but do not touch it. Go into each toilet stall, lock the door, and crawl out underneath. Leave the bathroom fast.

FOR YOUR MOTHER TO FINISH TALKING

Pull on her sleeve. When she says, "In a minute, dear," stand between her and her friend so that, in order for them to see each other while conversing, they have to look over your head or around you. As a result of this interference, your mom will say, "How about a cookie, dear," and produce a package of Mallomars from her tote bag. Sit on the couch and consume: Bite off the graham-cracker bottom, peel off the chocolate, piece by piece, and then play with the marshmallow center, pulling it like taffy. Pretend it is Play-doh until fingers are sticky and the marshmallow has turned an off-white from your dirty hands. Pop entire thing in mouth. Do not wash sticky hands, but see if you can pick hairs of wool off the couch with them. Announce that you are bored. Ask if it is time to leave yet, how much longer, what's for dinner, and can you please wait in the car. Do not say good-bye. Run out the door, hop into the front seat, open the glove compartment, and see what's new.

TO BE EXCUSED FROM THE DINING-ROOM TABLE

Say that you are stuffed. Tilt backward until chair rests precariously on its two back legs. If your parents try to make you eat one more bite, threaten to throw up. Return chair to upright after being asked.

Sink down until nose is level with the top of the table. Disappear altogether and steal napkins from everyone's laps.

Exhale real loud. Ask why you should have to wait while the grown-ups drink coffee.

Stretch.

Examine the salt and pepper shakers.

Make your napkin into a cootie catcher.

Walk your knife and fork around on the table as if they were people.

Rest chin on table. Stick chin in milk glass.

Jiggle leg until your mother asks you to stop because the table is shaking.

Lean back until your chair rests precariously on its two back legs. Fall over.

YANKEES

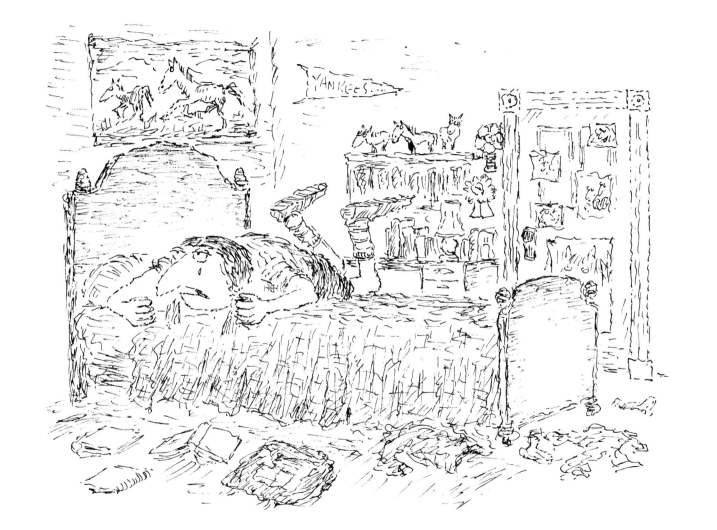

How to Act after Being Sent to Your Room

Slam the door.

"I hate them!

"I hate them I hate them I hate them I hate them! I could kill them! I wish they would drop dead! *I hate them!* I wish their heads would fall off! I wish they would get run over by a truck and never get up! This is *my parents!"*—grab doll, pull its head sideways, smash it to the ground. Kick it. "I'll get them. They'll see. I'll get them and they'll be sorry."

Throw yourself on the bed diagonally—legs dangling off, head buried in pillow. "I get blamed for everything! Every single thing that happens in this creepy house is my fault! *It's not fair!* Me! *Always me!* What did I do to deserve this? Nothing! Did I ask to be born? Did I ask them to have me? Did I? I *hate* being the oldest! *I hate it I hate it I hate it! The oldest always gets it—that's the story of my life!*

"Oh who cares anyway. I wanted to go to my room. I'm glad I'm here. That's where I wanted to be."

Pick up the cat and hold it on your lap. Hug it. Lean over and rub your hair in its fur. "I love you. I love you so much. I love you, silly cat, cute cat, pretty cat. I love you and not anybody else. I love you I love you I love you I love you I love you. You're the only one in the family that I love. You're the only one who understands me." Hug it tighter.

"Nobody else understands me. Nobody. Everyone else is against me. Everyone but you. My mother doesn't care about me. She really doesn't—I know. She doesn't love me. She doesn't listen to me. I tried to tell her. I tried to explain, but does she care? She never wants to hear my side. Never! It's always her side. Her side's the right side; her side's the one! Anyone else's side? For-get it! She thinks she knows everything. Well, she doesn't! *I hate her*—BIG UGLY KNOW-IT-ALL. FATHEAD. JERK. I always get blamed! I always get picked on! I always get it! *It's not fair!*

"Just wait. Just wait—they'll regret it! I'm never ever ever ever speaking to them again as long as I ever live—even if they speak to me first, even if they beg me: Just let them try. They'll see. I've had it. I'll give them the silent treatment for the rest of their lives!

"Boy if I die will they be sorry."

Open the door and stand at the top of the stairs. Yell, "Can I come down yet?"

How to Celebrate Christmas

BUYING GIFTS

Empty your bank. Make a pile of pennies, a pile of dimes, and a pile of quarters. Count the money. See how high you can stack the pennies before they fall over.

After having thought seriously about what your parents would like to receive, go to the five-and-ten-cent store and buy your mother a gold pin of a bird. Buy your father a scale model of the lunar module.

At dinner that night, say, "Hey, Dad, I bet you can't guess what I got you. Wanna know?" He won't want to know—he would rather be surprised—so whisper what you got him real loud in your mother's ear. If she can't understand, take her into the kitchen and tell her again. Go back and sit down, put your elbow on the table, rest your head on your hand, and give Dad a big grin. Tell him that he'll love it, it's real neat, and wouldn't he just like a little hint?

After dinner, drag your brother into your room, close the door, and show him what you bought, unwrapping the gift so that you can reuse the packaging: Take off the ribbon without untying it; try to sneak off the Scotch tape so the wrapping paper doesn't get torn or bruised, but do not succeed; open one end only and slide the present out. The wrapping paper should still be shaped like a box. After your brother admires the gift, slide the gift back in, get impatient, jam it at the corner, and rip the paper.

Lie on the floor on your back with your arms crossed behind your head and think about Christmas. Wonder what your presents will be. Wish that Christmas were tomorrow.

IN THE CLOSET

Turn head sideways, then upside down to read tags. Discover that your brother is getting that large present in back. Try to find a present for yourself that is equally large. Shake some gifts; hold others up to light to try to see the package through the wrapping paper. You hear someone coming. Quick—close the door and throw yourself on the couch. Listen. As soon as you realize that you were imagining things, open the closet and continue.

Later: When your mother asks if you were in the closet, be very busy making ten perfectly symmetrical rows of Fritos on the rug. Do not look up. Say, "No, Mom. Why?" She will say that someone was sneaking a look at the presents. Adjust a Frito while you suggest that it was probably your brother. If she questions you further, look up, stare her straight in the eye, and insist that you would never do anything like that—it would ruin Christmas. In fact, come to think of it, you are sure it was your brother—you saw him coming from that direction earlier in the day.

STRINGING GARLANDS

Eat the popcorn. Squeeze the cranberries to get your fingers red. Put red measle polka dots all over your face. Tell your mom to come fast, you're sick, and when she looks in, giggle.

PERFORMANCES

Dress as Mary with a blue beach towel for a veil. Dress your brother as Joseph in your dad's bathrobe. Cast the dog as Jesus. Abandon the act because the dog refuses to stay on its back, covered with a blanket, in a cardboard box.

DECORATING THE TREE

After your dad saws off the top of the tree so it fits into the living room, help spread the lights on the floor to untangle them. Your father will be giving everyone orders, trying to run the show. When he gets frustrated and starts swearing, explain to him that *you* know how to untangle them. He will get furious and tell you to shut up. Cry—this will make him feel bad so he will apolo-

61

gize and tell you that he just lost his temper, but you should still be quiet and not think that you know everything. When it's time to test for blown-out bulbs, yell, "Don't anybody move, I'll do it," and plug in the light cord. Then yell, "Don't let the dog step on the lights," run over to stop him, and trip over the lights.

Argue about who gets to hang the tiny angel with silver wings. Insist that your brother is not old enough to be entrusted with the responsibility.

Throw tinsel at the tree after you have hung the fancy bulbs and got bored hanging the plain ones. The minute she sees you throwing tinsel, your mom will say that you should hang it, not throw it, and give a demonstration, illustrating that tinsel should be hung strand by strand and never in clumps. Appear to pay attention, but secretly look at your brother and make a face. Hang tinsel until she goes to get the vacuum cleaner; then immediately start throwing it at the highest branches. If she notices when she returns, either say that you had to throw it because you couldn't reach, or deny it.

THE NIGHT BEFORE

Get into bed with your entire collection of stuffed animals because it is a special night. Tell yourself that the sooner you fall asleep, the sooner it will be Christmas. Lie on your side. Curl up into a ball. Turn over on your other side. Wait for sleep. Turn over on your back. Keeping your body absolutely still, wiggle your toes. Stare at the ceiling, pull your hands down over your face, and think, "I can't stand it, I'll never be able to wait." Smile. Cover your mouth to hide the smile even though no one can see it because the room is dark except for your brother's night-light.

Ask your brother if he is up and tell him that you can't sleep. To keep him awake, teach him something he has wanted to learn for a long time, like how to whistle or burp on demand. Then inform him that there is no Santa Claus and that if he sneaks into the hall and looks, he will see who really puts out the presents.

Turn over on your stomach. Turn over on your side. Curl up. Say to yourself, "Please let me fall asleep. Please, please, please, let me fall asleep. I'll never fall asleep." Turn over on your other side. Sigh. Wake up—it's Christmas, and if your parents said you could wake them at 7:00 A.M., it's 5:00 A.M.; if they said 8:00, it's 6:00.

OPENING PRESENTS

Ignoring all instructions to save ribbon, and being certain that all gifts become separated from their cards, open presents one of these ways:

- Save the best for last: Work your way from small to large.
- Space out the good presents: Alternate a few small with a large; end with the largest.
- Grab the first thing you see. Continue willy-nilly.
- Immediately open the largest presents; then get jealous of your brother, who still has large ones to go.

When you unwrap something you especially like, say, "Mom, Mom, Mom, Mom, Mom, Mom," to get attention. As soon as she says, "Very nice, dear," throw it aside and start on the next present.

When you unwrap something that you don't like, discard it quickly. Your mom will inevitably notice and say, "What was that, dear, I didn't get to see it," and you will have to hold it up.

When your father wants to snap your picture but cannot get the camera ready in time to catch the moment you open the box, put the gift back in the box, take it out as though you had never seen it, look up and smile.

ENJOYING YOUR GIFTS

After you have finished unwrapping, look around. Notice that your brother is still at it. Check his pile, then yours—does he work slower or did he receive more? Your parents will notice your concern and say, "Yes, your brother got two more presents than you, but we got you the guitar and that cost a lot of money, so it really comes out the same."

Call up your best friend and tell him everything you got and listen while he tells you everything he got. Roughly estimate the number of gifts you received, including stocking stuffers, and if he estimates higher, revise estimate upward. Claim that you received at least twenty-five dollars' worth of presents. When he replies that he got at least fifty dollars' worth of presents, say, "Well, guitars cost a lot, so I probably got at least a hundred dollars' worth of presents." When your little brother overhears and says, "I'm going to tell Mom that you were bragging about money," tell him to shut up.

How to Torture Your Sister

She ate her jelly doughnut at lunch. You saved yours. It is now two hours later:

Sit down next to your sister on the couch. Put the jelly doughnut on a napkin in your lap. Leave it, untouched, until she asks you if you still want it. Then begin eating: "Mmmmmmmmmmmmmmmmmm. This is soooooooo good." Take a large bite and chew with mouth open so she gets a good view. Swallow and run tongue over lips. "Mmmmmmmmmmmmmmmm." Stick tongue in jelly center and wave it around in the air before pulling it back in mouth. "Don't you wish you had some?" Take tiny bites. Lick fingers in between. "Boy—there's nothing like having a jelly doughnut in the middle of the afternoon!" Pop last bite in mouth and pat stomach.

Wander into the room when she calls a friend on the telephone. Pick up a book and sit down on the couch. Pretend to read. Then mimic her as she begins her telephone conversation:

Hi, how are you? *Hi, how are you.* Wha'd you do today? *Wha'd you do today?* What? Wait a minute, my sister's driving me crazy. *Wait a minute, my sister's driving me crazy.* Would you cut it out. *Would you cut it out.* You dirty creep. *You dirty creep.* Stop repeating me! *Stop repeating me!* I'll kill you if you don't stop! *I'll kill you if you don't stop!* I said STOP! *I said STOP!* STOP IT!!!! *STOP IT!!!!*
Put down book and run.

She is eating peanuts. Whisper in her ear, "You can turn into an elephant if you eat too many peanuts. I read it in the *World Book.*"

Pretend to eat shaving cream:
"Mmmmmmmmmmmmm. This ice cream is soooooo good. Wanna try some?"

Follow her everywhere.

Imitate her best friend talking. Say that her best friend is fat.

Talk to your mother while your sister is listening:
"Do you remember Christmas when I was three years old and you gave me that stuffed animal? That was so much fun." Turn to your sister: "You weren't alive."

70

You are in bed with the flu, watching television. She has been told to keep out of your room so that she doesn't catch it, too. As she walks by the door, stare goggle-eyed at the TV:

"Oh my goodness! That's incredible! I've never seen anything like this in my life! I can't believe it! Wait till I tell the kids at school." Do not remove eyes from set, staring in amazement. "I wouldn't miss this for anything! I really don't believe it." Look at your sister. "What?" Move over on the bed. "Of course there's room for you."

Check hallway to see if coast is clear. Pull her into your room, close door, put finger to lips, and speak in conspiratorial voice:

"I've got to tell you something. You're adopted! No kidding! Honest. Dad showed me the papers." Pause, scrutinizing her face. "You know, now that I look at you, I can tell. You really do look different from the rest of us. I mean, didn't you suspect it your-self? Dad even knows who your real parents are—they live in New Jersey—but he said he's not going to tell you anything until you're older. He swore me to secrecy. I'm just telling you because I think you ought to know, but if you tell him I told you, I'll kill you."

You are eating Jell-O. She is sitting next to you at the table:

"By the way, did you know that Jell-O's alive? Seriously. See how it wiggles?" Jiggle bowl. "I'm telling the truth, I really am.

Lookit, you've heard of jellyfish, haven't you? Need I say more? Jell-O's like jellyfish, only you eat it." Move bowl very close to sister and jiggle again. "If it doesn't eat you first . . ."

She is watching television. You are watching her:
"It's too bad about your lower lip. You've noticed, haven't you? You're kidding. Come here, I'll show you." Take her into the bathroom and place her squarely in front of the mirror. "See? It's amazing, isn't it—your lower lip looks exactly like a frankfurter. I can't believe you've never noticed. It's so obvious." Shake head despairingly. "Your looks would be just perfect otherwise. Here, I'll show you." Put one finger at each end of her bottom lip and push down. "There! Now you're normal!"

The lights are out. The two of you are lying in your beds:
"There's an invisible man who lives under your bed. If you did anything wrong today, he'll get you in the middle of the night. Oh yes you did! Don't you remember that thing Mom got mad at you for? Good night."

How to Ride in a Car

Insist that you don't have to go to the bathroom. If your mother points out that sometimes you think you don't have to go and then it turns out that you do, remember last time, say that you are sure. When she suggests that you try anyway, claim that you just went. Then have this conversation with your sister:

"Dibs on the front seat."

"I want it. You always get it."

"You had it last time."

"*You* had it last time."

"I did not."

"You did too."

"Did not."

"Did too."

"Did not."

"Did too."

"Did not. What a liar."

"I am not a liar. Mom, he called me a liar. Mo-om!"

"I did not."

73

"Did too. Mom, I want the front seat; he had it last time. I get it, don't I?"

Your mother will say that she can't stand it another minute—no one is sitting in front. Then she will finish making the peanut-butter-and-butter sandwiches and cream-cheese-and-grape-jelly sandwiches. Sit on the hood above the headlights, waiting to leave.

When your sister climbs up and sits next to you, call her a copycat. Keep repeating, "Copycat, copycat, you are a copycat. Copycat, copycat, you are a copycat." Then announce that you intend to sit next to the window. That's no problem, your mother will explain, since there are two windows—you and your sister can each have one. Discuss windows with your sister:

"I want that window."

"I want it."

"You got it last time."

"You got it."

"I did not."

"You did too."

"Did not! You make me barf. I bet you eat worms."

"Mo-om! He said I eat worms."

Now your mother will point out that since your sister knows she doesn't eat worms, why is she worried about it. Furthermore, if she hears one more word, neither of you is going anywhere. As you get into the car, mumble, "Good, I don't want to go." Then tell your sister to stay on her own side. Show her where her side

begins: Draw an imaginary line down the back of the front seat, across the floor, and up the back seat. Tell her not to cross the line. Tell her to stay where she belongs. Say, "Stay on your own side—I don't want your cooties."

Shortly after the car leaves the driveway, announce that you have to go to the bathroom and deny kicking your sister. Agree to hold it for a while, and when she kicks you back, kick her again. While she cries, think "Crybaby." Wish that she would fall out the window. Wish that she were kidnapped and that your parents couldn't afford to buy her back and that there was absolutely nothing that anyone could do about it. Explain to your mom that you wouldn't have kicked your sister if she hadn't kicked you first. Your mother will say, "Cut it out, both of you, or I'll stop the car right here." Then she will slow down the car and ask if that's what you want. After that, she will say that you should each stay on your own side, that there's more than enough room for each of you.

Read a comic. Pay no attention when your mother points out interesting things to see.

Ask if you are almost there yet.

Play out-of-state and, in the middle of arguing about who called out-of-state first, declare that you have to go real bad. As soon as the car pulls into a gas station, run, holding your crotch, and just make it. Ask to stop for ice cream.

Good grief—your sister thinks she is getting sick. Make a horrible face. Say, "Euuuuuuuuuuu, P.U., not again." Holding your nose so that you sound as if you were underwater, say that the car is going to smell dis-gusting. Say "Yuck," "Get me out of here," and "Don't throw up on me." Tell your mom that, frankly, you think it's stupid to take the creep anywhere, since she practically barfs the minute she gets in the car. To the creep, say, "Want a sandwich? Hee-hee-hee." When she clutches her stomach and tells your mom to make you stop, say, "All I said was, 'Want a sandwich?'"

Prop up your feet on the window ledge so that people driving by can see them. Ask to stop for ice cream.

Ask if you could spend the night at a motel with a color TV and a swimming pool.

Tell your mother that she is driving over the speed limit. Tell her that her seat belt isn't fastened. Tell her that a person's always supposed to drive with both hands on the wheel. Tell her that she really shouldn't smoke; it's bad for her.

Ask if you are almost there yet.

"Ninety-nine bottles of beer on the wall. Ninety-nine bottles of beer. If one of those bottles should happen to fall, ninety-eight bottles of beer on the wall. Ninety-eight bottles of beer on the wall. Ninety-eight bottles of beer. If one of those bottles should happen to fall, ninety-seven bottles of beer on the wall. Ninety-seven bottles of beer on the wall. Ninety-seven bottles of beer. If one of those bottles should happen to fall, ninety-six bottles of beer on the wall. Ninety-six bottles of beer on the wall. Ninety-six bottles of beer. If one of those bottles should happen to fall, ninety-five bottles of beer on the wall. Ninety-five bottles of beer on the wall. Ninety-five bottles of beer. If one of those bottles should happen to fall, ninety-four bottles of beer on the wall. Ninety-four bottles of beer on the wall. Ninety-four bottles of beer. If one of those bottles should happen to fall, ninety-three bottles of beer on the wall—"

"I can't stand it! Stop it! I can't stand it another minute!"

Deny that you are the person kicking the seat.

Be engrossed looking out the window while you casually move your hand on the seat over to your sister's side. If she doesn't notice, poke your pinky into her leg. Then she will say for sure, "You're on my side. Get off. Ma, he's on my side." Remove hand, say that you were not, look out window, put hand back, and leave it there until she hits you. Hit her back. Hit her again. Kick her while hitting her. Try pulling her hair. While she cries as she yells, "Stop hitting me," and hits you back, yell, "You hit me first," and hit her again. Kick the seat when you miss kicking her. Knock your head against the door and cry.

At this point, start listening to your mother as she finishes what she began saying after the first hit: "Stop it, both of you. I've had enough. You hear me? If either of you lays a finger on the other, if I hear one more word out of either of you, that's it. Understood? I'm never taking you anywhere again. Is that what you want? Is it? I swear—if I hear one more peep, I'm turning around right now."

Sniffle. Wipe your nose on the back of your hand. Sniffle again.

Lean back against the seat. Sniffle. Close your eyes: You are now riding in the back seat of a chauffeur-driven limousine with a TV—all alone. Think about your sister. Open your eyes and look at her. She is so disgusting. You hate her guts. Swear that you'll never talk to her again, never, even if she's dying. More than anything in the world, wish that you were an only child. Sniffle. Wipe your nose on your sleeve. Look down—there are the crusts of bread from your sandwich on the floor where you dropped them. Look out the window. Wonder if you're almost there yet.

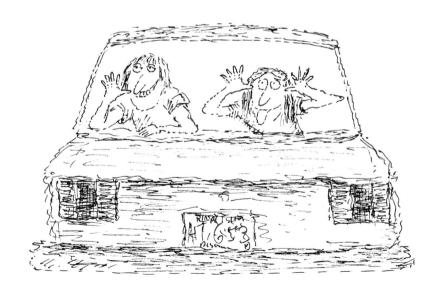

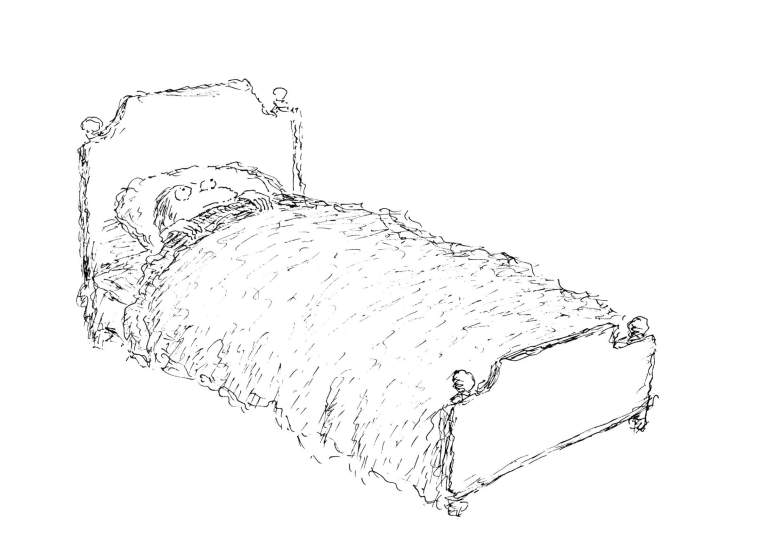

How to Sleep

Aw Mom, do I have to? I don't want to. But I'm not tired. Can't I just stay up a little while longer—just a teeny, weeny while longer? Oh Mom, come on. A little longer won't hurt. Just till I finish this chapter? Just till I finish this page?

Finish reading the page and start on the next one. Then start on the next. Read until your mother notices that you are still reading. As soon as she does, she will again inform you that it's time for bed. Say, "OK, but I just want to finish reading this sentence."

Finish the sentence. Finish the page. Keep reading. When your mom comes back into the room to find out what's holding you up, close the book quickly and act offended: "OK, OK, take it easy. I said I was coming!" Center the book carefully on the coffee table. Start to walk away from the table and suddenly get superstitious—if you do not center the book perfectly, something bad will happen to you. Return to the table and adjust placement. Adjust again. Notice the cat.

Pick it up and cuddle it. Scratch it behind the ears. Your mom will yell at you to hurry up. Yell back, "I've got to say good night to the cat! Do you want me to go to bed without saying good night to the cat? Honestly!"

As you leave the room, go very slowly. Stop at a table. Walk your fingers across the top of it. Spin a ceramic egg. Examine an ashtray. Answer your mom: "What? What? Yeah, just a second, I'm looking at this." Hold the ashtray up to the light and squint through the glass. Turn the ashtray over. See if the cat will walk around with an ashtray on its back. Try to balance the ashtray on top of the egg. Scream, "OK!!!!!!! I'm coming! Can't a person just look at something for a change?"

Kiss your mother good night.

"Mom, do you realize that I'm the only person in my entire class that has to go to sleep at eight-thirty? The only one!" Your mom will inquire, "If everyone in your class jumped off the Empire State Building, would you do it, too?" Then she will add that she has heard enough today already and you know what she means. As long as you are living in her house, you will do things her way. When you have your own house, you can stay up as late as you want.

Do not say anything, but make a horrible face when your mother turns her back, and erase it the second she turns around again. Wait until the absolute last second and maybe let her catch a glimpse. Wish that your parents had gone out tonight. If they had, you could lie to the baby-sitter and claim that your bedtime is later than it is.

As you climb the stairs to your room, stand sideways and skip each third step. Put on your pajamas, mismatching the buttons and the buttonholes. Forget to brush your teeth.

Run across the room and leap into bed so that nothing under the bed gets you. Lie under the covers awhile, just staring. Pick your nose. Examine what you dig out and scrape it off under the side table next to the bed. Then, leaning over so that your head and half your body are hanging over the side, pull out the horror comic that you have hidden underneath it. When you hear your mom coming in to check on you, stick the comic under the covers, grab a book off the side table, and pretend to read it. Tell your mom that you'll go to bed as soon as you finish reading this page. She will ask if you brushed your teeth. Say yes. As soon as she leaves, pull out the comic and continue reading. When she yells from downstairs, assure her that the light *is* out, and continue reading. Finish the comic; stick it under the pillow.

Turn out the light.

Lie in bed. Let your eyes adjust to the darkness. As cars pass by on the street, watch their headlights shine on the wall of your room and then disappear. Listen to the sound of your parents and their friends talking in the living room. Wish that you were with them. Get out of bed, sneak halfway down the staircase, sit on a step and peek at the company through the banister. Maybe your mother or father will notice you sitting there. Then you will get to appear in your pajamas and say hello to everybody. Go back to bed. Take your flashlight in bed with you.

Turn it on and cover the light with the palm of your hand. Watch your hand light up and turn pink. Get out of bed. Go to the bathroom and look in the mirror. Put the light under your chin and watch your chin light up. Open your mouth wide. Stick the lighted end of the flashlight in your mouth and watch your cheeks light up. Turn the flashlight off and on a few times to impersonate a neon sign. Send a light message in Morse code.

Run back to bed—you think that you hear someone coming. You were wrong. After waiting a few minutes to be sure—counting patterns in the wallpaper—pull the covers over your head and read a book under the covers with the flashlight. At the end of one page, you will be roasting hot. Throw off the covers, open your mouth, pant several times, and fan your face.

Yell downstairs, "Mom, I can't sleep. This blanket itches."

Try to be invisible by lying as absolutely flat as you can just in case a burglar comes into the room. Tuck your hands under you. Reconsider the situation: Perhaps if you scrunch into a ball, you will be less noticeable. Try it. Worry that there really is a creature from the center of the earth hiding in your clothes closet like your older sister said. Study the light fixture on the ceiling and imagine that it is a giant tarantula. Ask for a drink of water. When your mother brings it, inquire casually, by the way, has she ever seen a tarantula? If she says yes, once when she was digging in the backyard, decide to keep watch all night and not under any condition to fall asleep. Then fall asleep while keeping watch. If she says no, or only when she was at the Grand Canyon, fall asleep.

Fall out of bed and don't wake up.

How to Say Your Prayers

Our Father, who art in heaven, hallowed be Thy name. Thy kingdom come, Thy will be done, on earth as it is in heaven. Give us this day our daily bread, and forgive us our trespasses as we forgive those who trespass against us. And lead us not into Penn Station, but deliver us from evil, for Thine is the kingdom and the power and the glory for ever and ever.

I wish that there would be no war between anyone anywhere.

I wish that Grandma would get well.

I wish for a million dollars.

Amen.